U.S. ECM Aircraft

Written by Ken Neubeck

In Action®

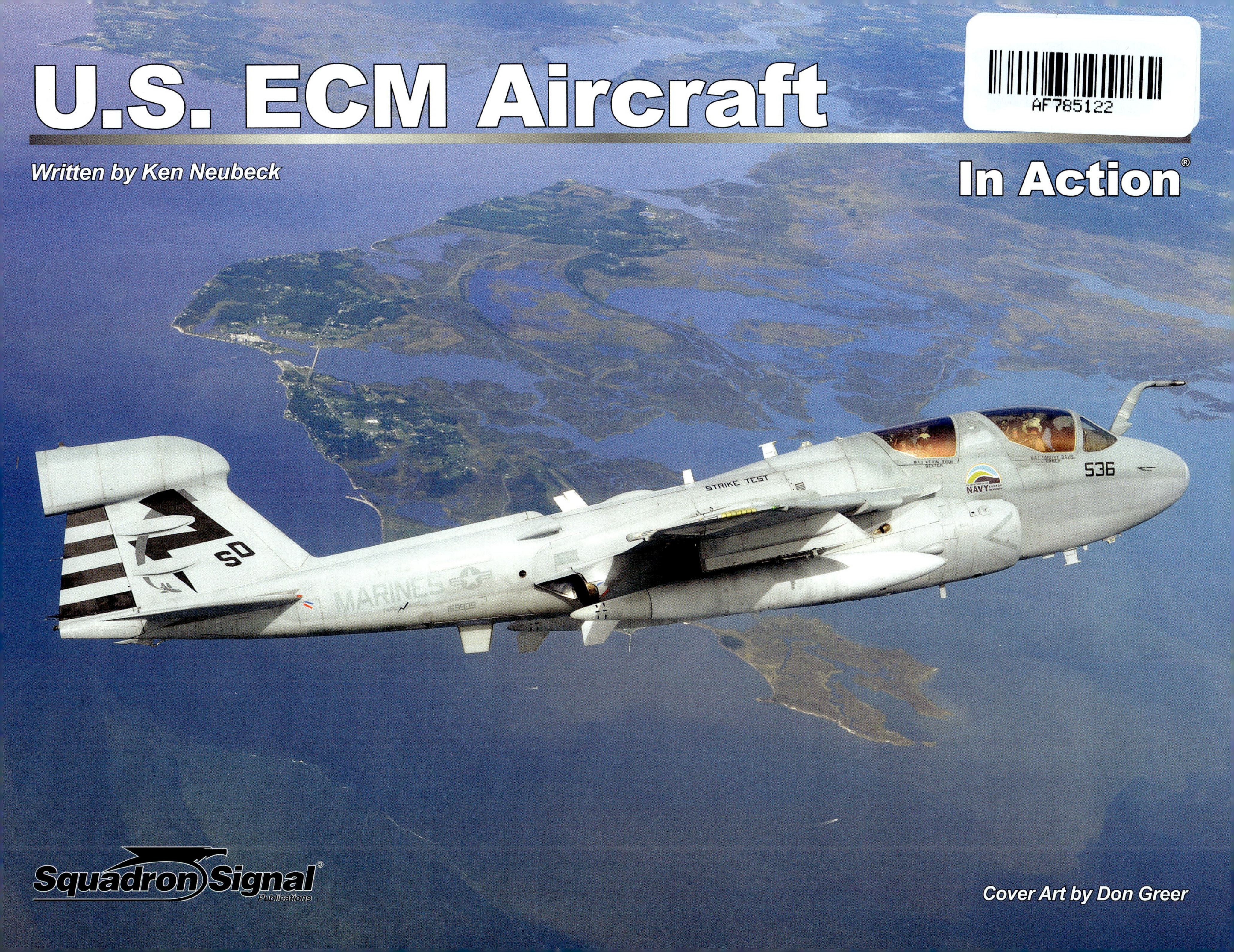

Squadron/Signal Publications

Cover Art by Don Greer

(Front Cover) An EA-18G, the newest ECM aircraft in the U.S. Navy inventory, lands on an aircraft carrier in 2013.

(Back Cover) An EA-6B Prowler from the "Scorpions," Electronic Attack Squadron 132, flies over the aircraft carrier USS *John F. Kennedy* in the Atlantic Ocean in 2004.

About the In Action® Series

In Action® books, despite the title of the genre, are books that trace the development of a single type of aircraft, armored vehicle, or ship from prototype to the final production variant. Experimental or "one-off" variants can also be included. Our first *In Action®* book was printed in 1971.

ISBN 978-0-89747-742-0

Proudly printed in the U.S.A.

1115 Crowley Drive, Carrollton, TX 75006-1312 U.S.A.

Military/Combat Photographs and Snapshots

If you have any photos of aircraft, armor, soldiers, or ships of any nation, particularly wartime snapshots, please share them with us and help make Squadron/Signal's books all the more interesting and complete in the future. Any photograph sent to us will be copied and returned as requested. Electronic images are preferred. The donor will be fully credited for any photos used. Please send them to:

Squadron/Signal Publications
1115 Crowley Drive
Carrollton, TX 75006-1312 U.S.A.
www.Squadron.com

(Title Page) An EA-6B Prowler aircraft conducts a test flight over the Southern Maryland coastline. It tested a biofuel blend of JP-5 aviation fuel and camelina oil. The Prowler successfully completed its inaugural biofuel flight to continue the Navy's goal of energy independence. The SD tail code and lightning motif markings indicate it is from the "Salty Dogs" of Air Test and Evaluation Squadron 23 (VX-23) based at Naval Air Station (NAS) Patuxent River in Maryland. (Kelly Schindler / U.S. Navy)

Acknowledgments

The author wishes to thank Larry Feliu, John Gourley, Jamie Hunter, Dennis R. Jenkins, Del Laughery, Ben Peck, Sal Picattagio, Josh Stoff, Cradle of Aviation Museum, Northrop Grumman History Center, U.S. Navy, U.S. Air Force (USAF), U.S. Marine Corps (USMC), National Archives, and the J.W. Hawkins Collection for their help in this project. Additional assistance was provided by Ken Katz.

Introduction

The art of Electronic Countermeasures (ECM) traces its roots to as early as 1904 during the Russo-Japanese War. During the war, land- and ship-based Russian spark-gap signal wireless stations successfully jammed wireless communications with senseless noise between Japanese battleships during an attempted attack on a Russian Navy base. ECM continued through World War I when British and German radio stations interfered with each other's signals throughout the Western Front.

ECM grew more sophisticated during World War II as it included many new types of countermeasures, which was particularly true with the advent of radar during the war. New countermeasures included dropping chaff to fool radar detection as well as other means to spoof navigation signals. Both British and German engineers developed radar defenses to deal with the opposing fighter forces. During the war, RAF Defiant aircraft carried high-power Mandrel jamming equipment to disrupt the operation of Germany's acquisition radars.

ECM interference techniques are accomplished by jamming transmission signals on specific radar frequencies to produce noise levels sufficient to hide echoes created by the aircraft signatures. The jammers' continuous transmissions provide a clear direction to the enemy radar, but do not provide range information or distance. In addition to incorporating radar-jamming equipment into the aircraft equipment bays, there was a need for external jamming pods to be hung from the weapons pylons of the aircraft wings.

ECM-equipped aircraft use three types of jamming techniques. These jamming techniques are spot, barrage and sweep. Spot jamming involves a narrowly-directed signal, but it can miss other locations. Barrage jamming simultaneously jams many frequencies, however this type of jamming utilizes a lower effective power and reduces range. Sweep jamming involves many signals sent across a frequency band, but a jammer's full power alternates across several frequencies within the band. While multiple frequencies are jammed in quick succession, it is not done simultaneously and thereby limits sweep jamming's effectiveness.

The primary difference between barrage and sweep jamming is displayed in the modulation techniques along with the size of the frequency band covered. Barrage jamming generally uses an amplitude-modulated (AM) signal that covers about 10 percent of the frequency bandwidth. Sweep jamming generally uses a frequency-modulated (FM) signal and the frequency is moved or swept back and forth over a very wide bandwidth. Practical limitations and difficulties are experienced when operators try to match the jammer frequency exactly to the radar's radiation frequency while it is necessary to broaden the bandwidth to cover all radars with frequencies in that band.

Spot jammers attempt to closely match a particular radar frequency. Importantly, when the jammer bandwidth is broadened, jammers require more power to effectively jam the radar as opposed to an exactly-matched jammer.

While electronic countermeasures are used in sea, air and land military equipment, modern military aircraft are the primary platform in the ECM battle because they can scan a larger patch of earth than sea or land based units. Effective use of ECM protects aircraft from enemy search radars as well as surface-to-air missiles or air-to-air missiles.

F3D-2Q 124603 is chocked on a hardstand in the 1950s. It flew with VFAW-3 (All Weather Fighter Squadron 3) from 1953 until 1958 out of Norton AFB in San Bernardino, California. During the late 1950s, the U.S. Navy modified 35 F3D Skyknight aircraft into ECM F3D-2Q, later the EF-10B, versions that would serve into the 1960s. (J.W. Hawkins collection)

An early implementation of ECM equipment into U.S. aircraft occurred in the late 1950s when the U.S. Navy installed equipment into the fuselage as well as two jamming pods on the wing pylons of an Douglas F3D Skyknight aircraft. This modified aircraft, later named EF-10B, was the first aircraft to detect Communist Chinese radar in 1958 during a Taiwan Strait Crisis and also detected Soviet ground radar in Cuba in 1962 during the Cuban Missile Crisis. It would later started Vietnam service in 1965 on an emergency basis to provide ECM protection for strike aircraft in limited numbers.

In 1959, the U.S. Air Force was involved with ECM development as well. Several obsolete RB-57A Canberra bombers were converted by Martin for use in air defense interceptor training. The aircraft had an extensive suite of ECM equipment installed throughout the aircraft – primarily in the bomb bay. An electronic warfare officer (EWO) operated all ECM equipment and replaced the photo-navigator used in the RB-57A.

However, it was the electronic attack threat during the Vietnam War that accelerated the research and development of effective jamming systems for the field of electronic countermeasures.

In July of 1964, U.S. aircraft first came under attack from surface-to-air (SAM) missiles during bombing raids near Hanoi. This action required the U.S. Air Force to simultaneously look at two methods to deal with this new threat. The first method was defensive jamming pods carried by fighter aircraft. The second method utilized a dedicated defense suppression aircraft that could identify the threat, warn other aircraft in the area, and attack against the threats.

An EA-1F Skyraider is displayed at the National Naval Museum at Pensacola, Florida. The Navy converted 54 Douglas A-1 Skyraider aircraft into the ECM version. The ECM version were known as the EA-1F Skyraider. An external jamming pod was added under a wing to handle ECM functions, but the pods are absent from all display and decommissioned aircraft. (John Gourley)

An EA-3B Skywarrior is displayed at the National Naval Museum at Pensacola. A total of 24 A-3A Skywarrior aircraft were converted into the ECM version EA-3B Skywarriors. These aircraft entered service in the late 1950s and saw action in Vietnam for the U.S. Navy. EA-3B later participated in Operation Desert Storm in 1991. Although removed for display, all ECM equipment was internally stored and not visible. (John Gourley)

U.S. ECM Aircraft During Vietnam

The United States participation in the Vietnam War precipitated major development of ECM-equipped aircraft to provide tactical defense penetration and suppression that continued into the modern era.

ECM jammer pods, such as the AN/ALQ-71 (A-Air Force, N-Navy/A-Airborne, L-Countermeasures, Q-Special) and the AN/ALQ-72, were attached to wing pylons on fighter aircraft such as the Republic F-105 Thunderchief. While these pods helped, early pods transmitted at low power and each aircraft was required to carry several pods to effectively jam electronics.

Additional ECM aircraft that saw action in Vietnam were the EA-1F Skyraider, the EA-3B Skywarrior, and the EF-10B Skyknight.

The EA-1F Skyraider, based on the A-1 aircraft, had a Wright R3350 radial engine and a crew of two (pilot and ECM operator). It was flown by the U.S. Navy, Marines and Air Force. The EA-3B Skywarrior, based on the Navy A-3 aircraft, sometimes doubled as a mid-air refueler. The EA-3B was a twin-engine jet aircraft with a crew of three, a pilot, a navigator and an ECM operator. All of these aircraft carried externally-mounted jamming pods attached to wing pylon stations to accomplish their ECM missions.

Eventually, ECM concepts were incorporated in jet aircraft. The Douglas F3D-2Q Skyknight, later redesignated as the EF-10B, was the first Navy jet aircraft dedicated to the Electronic Warfare (EW) mission. It replaced the propeller-driven Douglas AD-5N Skyraider, which was a four-seat night-attack version of the A-1. The EF-10Bs provided critical threat warning and radar jamming support for both the Navy and Air Force strike and reconnaissance operations over North Vietnam. The aircraft was retired in 1970.

Early ECM aircraft were all originally designed for other missions. Repurposed airframes resulted in ECM effectiveness limitations that were exposed during the Vietnam War when these aircraft challenged North Vietnam's Soviet-equipped air defense stations. Consequently, U.S. military planners realized that a "design from scratch" ECM aircraft was required rather than continued modifications to existing aircraft.

Scratch design allows for optimal placement of ECM components. Placement mattered for antennas and allowed advantages with heating, cooling and power sources located on the aircraft. In the interim, ECM jammer pods were attached to fighter and attack aircraft from the late 1960s through the Vietnam War.

Target ECM frequency ranges or "bands" as defined in the radio spectrum were:

Band	Frequency
1	64-150 MHz
2	150 – 270 MHz
3	270 – 500 MHz
4	500 – 1000 MHz
5/6	1.0 – 2.5 GHz
7	2.5 – 4.0 GHz
8	4.0 – 7.8 GHz
9	7.8 – 11 GHz

A Martin EB-57A Canberra tactical bomber flies on a mission. The EB-57A was an early Air Force effort into ECM aircraft. This aircraft was one of the first ECM aircraft used by U.S. Air Force during the Vietnam War in the suppression of SAM sites. (U.S. Air Force)

A Boeing EB-47E "Tee Town" Stratojet aircraft rests on a hardstand. One conversion program equipped 40 B-47E bombers with ECM jammers located inside the bomb bay along with traditional bombs. The pods were called "Tee Town" pods. (National Archives)

The Douglas EB-66 Destroyer was a B-66 light bomber equipped with a Quick Reaction Capabilities (QRC) 160 ECM pod and used briefly in the ECM role over Vietnam. It could only fly limited flight envelopes and had service limit vulnerabilities. (National Archives)

Gen. Gregory S. Martin tours an NKC-135. Two "Big Crow" ECM test aircraft were developed. The NC-135N was based on a Boeing C-135 Stratolifter, and the NKC-135 was based on the KC-135 Stratotanker. (Tech. Sgt. Robert J Horstman / USAF)

A two-seat F-105F Thunderchief awaits a mission in during the Vietnam War. It was modified with ECM equipment used for detection of North Vietnamese SAM radar sites during bombing runs. (National Archives)

An EA-6B returns from a mission. The EA-6B Prowler was another major turning point in ECM aircraft development. The Prowler was developed prior to retirement of the EA-6A Intruder and had expanded ECM capabilities. (Del Laughery)

Vietnam Era ECM-Dedicated Aircraft

A major turning point for ECM aircraft would begin in 1962 when the U.S. Marines awarded a contract to Grumman for a ECM version of the A2F Intruder attack aircraft. The program modified a prototype A2F aircraft, Bureau Number (BuNo) 148618, into an ECM version that was designated as the A2F-1H. Five additional A2F aircraft were modified into the A2F-1H configurations. Later, aircraft were modified for the ECM mission with the new designation of EA-6A Intruder. The new Intruders were fielded in 1968. The EA-6B followed. It was specifically built for ECM.

The original joint Air Force and Navy (AN) EA-6B aircraft had the AN/ALQ-99 ECM system that could operate in four Target ECM frequency bands: Band 1, Band 2, Band 4, and Band 7. After deployment in Vietnam, it became apparent that additional band coverage was needed to deal with Soviet-designed electronic warfare (EW) radars. Consequently, the next version of the EA-6B was the "Expanded Capacity" (EXCAP) version. The EXCAP added coverage for Bands 5, 6, 8, and 9. The EA-6B was equipped with both low band and high band AN/ALQ-99 pods.

The EA-6B entered service as part of a dedicated ECM squadron in July 1971. It saw action in Vietnam in 1972, where it conducted ECM missions for both the U.S. Marines and Navy. Afterward, the aircraft became the staple of U.S. ECM activity during combat. Fewer than 50 Prowlers remain in service as of 2016. All U.S. Navy Prowlers have been retired. There are only four Marine squadrons from Cherry Point in North Carolina with EA-6B Prowlers, but these will be retired by 2019.

Meanwhile, the U.S. Air Force addressed the immediate need for an ECM-dedicated aircraft. ECM aircraft clearly needed the same speed as the strike aircraft they escorted. Thereafter, the Air Force introduced fighter-style aircraft for the dedicated ECM aircraft role during Vietnam instead of modified bomber aircraft such as the EB-66, which was a modified B-66 light bomber. The North American F-100 Super Sabre aircraft was initially considered for this role and designated as Wild Weasel. However, the F-105D Thunderchief strike aircraft was faster than the F-100 escort. In 1966, the F-105F was designated as the strike aircraft escort.

The F-105F Thunderchief, Wild Weasel III, aircraft was equipped with a wing-pylon-mounted ECM pod along with an ECO in the rear seat to monitor radar signals. Once a SAM radar site was spotted, the ECO in the rear crew station would direct strike aircraft pilots to the site. The strike aircraft would attack the missile site with air-to-ground missiles. Tactics were developed such that Wild Weasel aircraft would precede strike flights, and would clear the target area of radar guided SAMs.

As the war progressed, the F-105F was upgraded to the F-105G ECM model. The F-105G had the AN/ALQ-105 pod electronics mounted to the fuselage and allowed one more weapon on a wing pylon. A total of 61 F-105F aircraft were upgraded to F-105G specifications. Beginning in 1965, the F-105F and G aircraft were so effective at identifying and destroying North Vietnamese SAM sites that the North Vietnamese often turned off SAM radars to avoid detection. This avoidance of ECM aircraft and resulting reluctance to use radar at the sites helped reduced Air Force aircraft losses.

Post-Vietnam Era ECM Aircraft

After the F-105 and F-4G Wild Weasel programs, the U.S. Air Force focused on dedicated ECM aircraft development with the General Dynamics–Grumman EF-111A Raven. The EF-111A used the AN/ALQ-99E ECM system. Like the EA-6B, it had an enlarged tail for the receiver. Instead of wing-mounted ECM pods, the EF-111A had a lower-fuselage mounted pod. Grumman got a contract to modify the F-111A Aardvark into the EF-111A with ECM equipment to begin in 1972.

The EF-111A Raven saw war service during Operation Desert Storm in Iraq in 1991 and Operation Deliberate Force in the former Yugoslavia in 1996.

Various studies were conducted that recommended that the EF-111 Raven be retired to reduce redundant aircraft with dissimilar maintenance needs. Ravens were retired in 1998. Consequently, four "expeditionary" EA-6B Prowler squadrons were established and composed of Navy and Air Force personnel to meet the needs of the Air Force.

The EA-6B became the primary ECM-dedicated aircraft for the U.S. Armed Forces for the next 10 years. It provided ECM support for all major American combat operations in the Mideast and Asia.

The EA-6B ECM system provides 360 degrees of receive and jam coverage, along with a tail warning sensor. Such a sophisticated system requires significant amounts of maintenance, repair and electronic component upgrades over time to keep it operational.

During this time, U.S. military aircraft employed alternate approaches to ECM protection. The Rockwell B-1B Lancer bomber has the AN/ALQ-161A ECM system onboard counteract threats. The system was developed by Airborne Instrument Laboratory (AIL) during the 1980s and consists of 107 Line Replaceable Units (LRU). The system is housed in several B-1B flight stations with passive antennas and high-power jamming transmitters positioned behind the wing root panels and inside the empennage bay.

The Northrop Grumman B-2 Spirit bomber has significant stealth features built into the aircraft. The skin is made of radar-absorbent material that blends into smooth contours to minimizes its return on radar. Its four engines are concealed inside the airframe to reduce infrared signature. It was first flown in 1989. The B-2's initial war action was in Bosnia. The B-2 is able to conduct bombing missions with the lowest amount of support aircraft because of its stealth ability. However, the B-2 was extremely expensive to build and maintain. Only 21 B-2 stealth bombers were built.

The AN/ALQ-135 ECM jamming system Tactical Electronic Warfare Suite (TEWS) on McDonnell Douglas F-15 Eagle and F-15E Strike Eagle aircraft is a similar defense. Similar to the B-1, this protection is accomplished through imbedded equipment. The system jams and tracks multiple anti-aircraft missiles in addition to other threats.

A number of U.S. fighter and attack aircraft faced electronic threats from potential enemy sources. As a result, the AN/ALQ-131 and the AN/ALQ-184 attachment pods were developed to mitigate threats.

Along with other modern fighters, the Lockheed Martin F-22 Raptor and Lockheed Martin F-35 Lightning II stealth fighters are both equipped with ECM hardware and capabilities for current and future warfare.

An EF-111A Raven is chocked on a hardstand. Beginning in 1972, the supersonic EF-111A Raven was another aircraft used for ECM missions. Ravens were modified from the F-111A Aardvark by Grumman. (John Gourley)

A B-1B Lancer banks over the desert during a mission. The B-1B bomber is outfitted with a complete ECM suite that allows the bomber to jam enemy radar during bombing missions. More than 60 B-1 Lancers remain in service currently. (U.S. Air Force)

A General Electric AN/ALQ-87 is mounted on a wing pylon of a McDonnell Douglas F-4 Phantom II. The AN/ALQ-101 jammer pod replaced the AN/ALQ-87. (John Gourley)

The Westinghouse AN/ALQ-101 ECM jammer pod was developed in 1966 and widely used on various U.S. Air Force aircraft during the Vietnam War such as the F-111, F-4, LTV A-7 Corsair II, and RF-4C photographic reconnaissance Phantoms. (John Gourley)

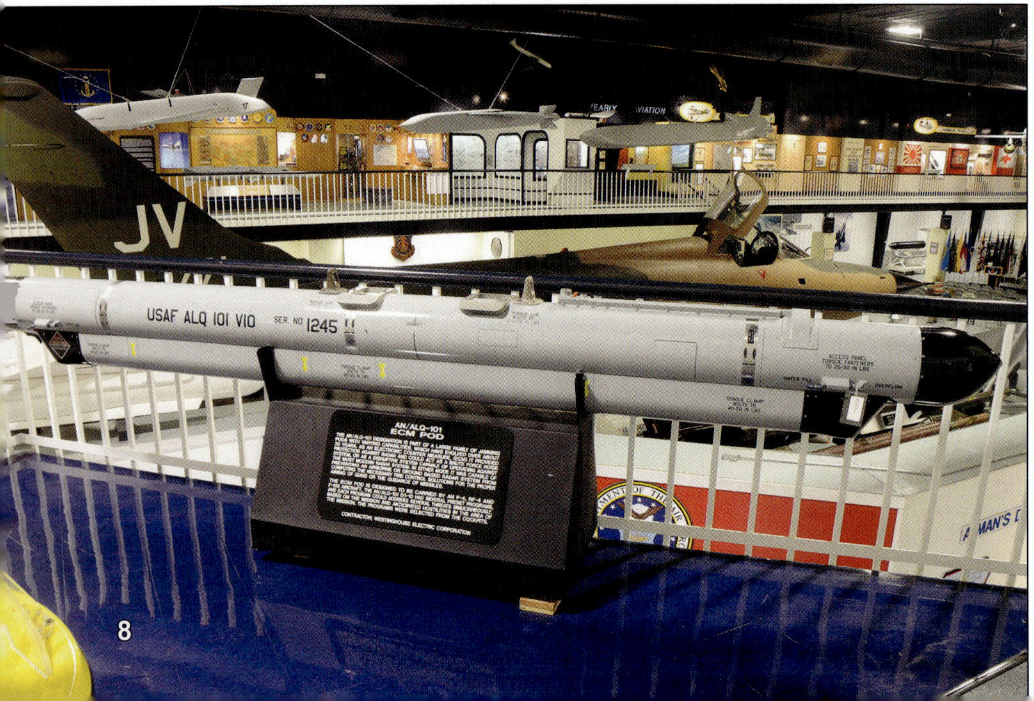

Early ECM Jammer Pods

Electronic countermeasure jammer pods were designed to be fitted to existing fighter or attack aircraft. The pods were either attached on wing pylons or on the lower fuselage centerline. These jammer pods are designed to deceive radar signal trackers and deny target information to enemy forces. This deception is accomplished through target pattern distortion or false target generation.

After development of the Wild Weasel jammer pods, there were subsequent jammer pod designs used on aircraft. Some of the newer pods include the General Electric AN/ALQ-87 and the Westinghouse AN/ALQ-101 ECM pods.

ECM jammer pods are additional structures mounted onto existing aircraft. There are two basic methods to mount these pods. For wing attachment pods, power and control signals are fed to the pod through a cable assembly located inside of the wing pylon that interfaces with connector located atop the pod. For lower-fuselage pods, the hookup is accomplished through a cable extending from the aircraft fuselage to the top of the pod. The harness connects with aircraft wiring and allows the pilot or ECO operator to electronically control the ECM jammer.

The AN/ALQ-87 was manufactured by General Electric and is a 1 to 8 GHz (Bands 5 through 8) FM barrage jamming pod. It was used on several Air Force aircraft such as the McDonnell F-4 Phantom, McDonnell F-101 Voodoo, Republic F-105 Thunderchief, General Dynamics F-111 Aardvark, Lockheed AC-130 Hercules gunship, and the Grumman OV-1D Mohawk light observation and attack aircraft during Vietnam. The pod was typically mounted and carried on one wing pylon of these aircraft. In 1966, this design became obsolete and was replaced by the AN/ALQ-101.

The AN/ALQ-101 jammer pod was a noise and repeater jammer that was used on several aircraft from 1966 through Operation Desert Storm in 1991. In contrast to the AN/ALQ-87 pod, the AN/ALQ-101 pod was three feet longer and was mounted on the lower centerline fuselage for Air Force aircraft such as the RF-4C, LTV A-7 Corsair II and the F-111. The RF-4C was the primary platform for this pod. The pod was initially called Sesame Seed and was designed for use in the 2 to 8 GHz range (Bands 7 and 8), but upgrades extended the upper-frequency coverage to 20 GHz. Four major upgrades happened over the pod's lifetime. Version 8 was the final version.

The AN/ALQ-101 pod saw more than 20 years of service from the Vietnam War into the Persian Gulf War. The final production version and the most numerous unit in service was the AN/ALQ-101(V) 8. This model was the first electronic countermeasures pod to use a gondola layout. The gondola layout includes a trough compartment underneath the main body. This location considerably increases the available volume in the pod without a significant increase the cross-sectional area.

Early designs such as the AN/ALQ-87 and AN/ALQ-101 worked sufficiently. However, they needed additional improvements to the effective power output as well as reliability performance and improved maintenance ease. These improvements were realized in the next iteration jammer pod, the AN/ALQ-119, which made the AN/ALQ-101 obsolete.

The Westinghouse AN/ALQ-119 ECM noise/deception pod replaced the AN/ALQ-87 and the AN/ALQ-101 pod. It was used on various fighter aircraft such as the F-4, F-100, F-101, F-105 and F-111 during the Vietnam War. (U.S. Air Force)

An F-111A Aardvark of the 48th Tactical Fighter Wing carries an AN/ALQ-119 ECM pod mounted on the lower fuselage as it flies above RAF Lakenheath in 1983. (U.S. Air Force)

An F-4E Phantom II from the 480th Tactical Fighter Squadron (TFS) carries an AN/ALQ-119 electronic countermeasures pod over Germany in August of 1982. The pod is located on the lower fuselage under the cockpit. (Master Sgt. Sutherland / U.S. Air Force)

A Lockheed AC-130 Hercules gunship carries an AN/ALQ-119 ECM pod on one of its wing pylons. (John Gourley)

A Fairchild Republic A-10 Thunderbolt II carries a Westinghouse AN/ALQ-131 ECM noise/deception jammer pod. The pod was also used on the A-7, F-4, F-15, General Dynamics F-16 Fighting Falcon, F-111, and C-130. The pod weighs 675 pounds. (John Gourley)

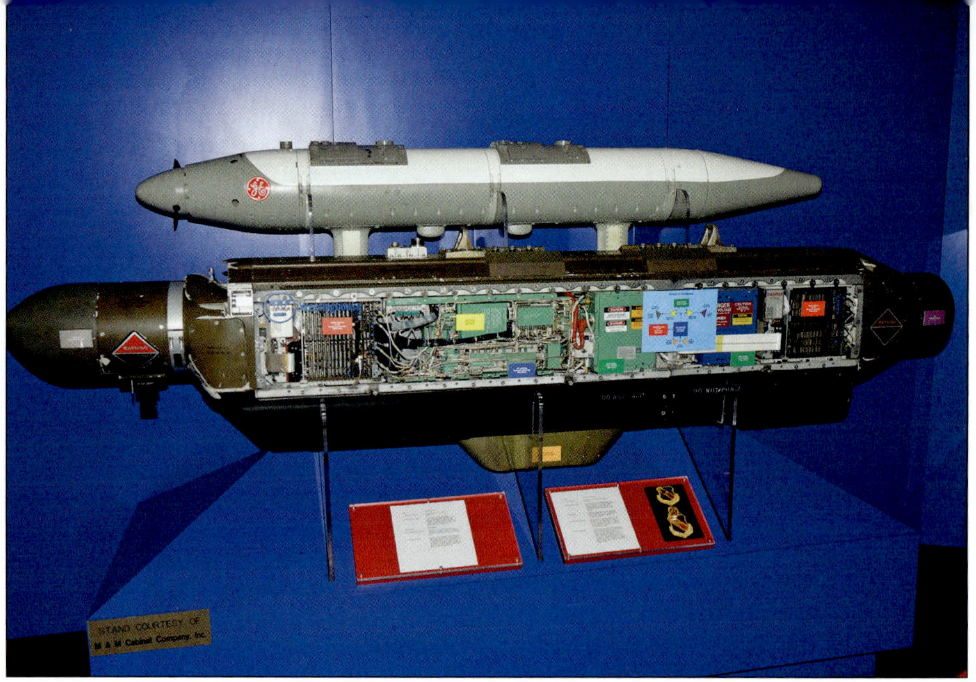

The AN/ALQ-131 ECM pod is more than nine feet long and displayed without a cover to show the tightly packaged modules. An AN/ALQ-87 ECM legacy pod is in the background. It was retired in the 1970s. (John Gourley)

The center section of the AN/ALQ-131 ECM pod is the Exciter section, which contains tightly packed electronic modules, wire harnesses and rigid cables. The tight proximity of components impacts pod maintenance. (John Gourley)

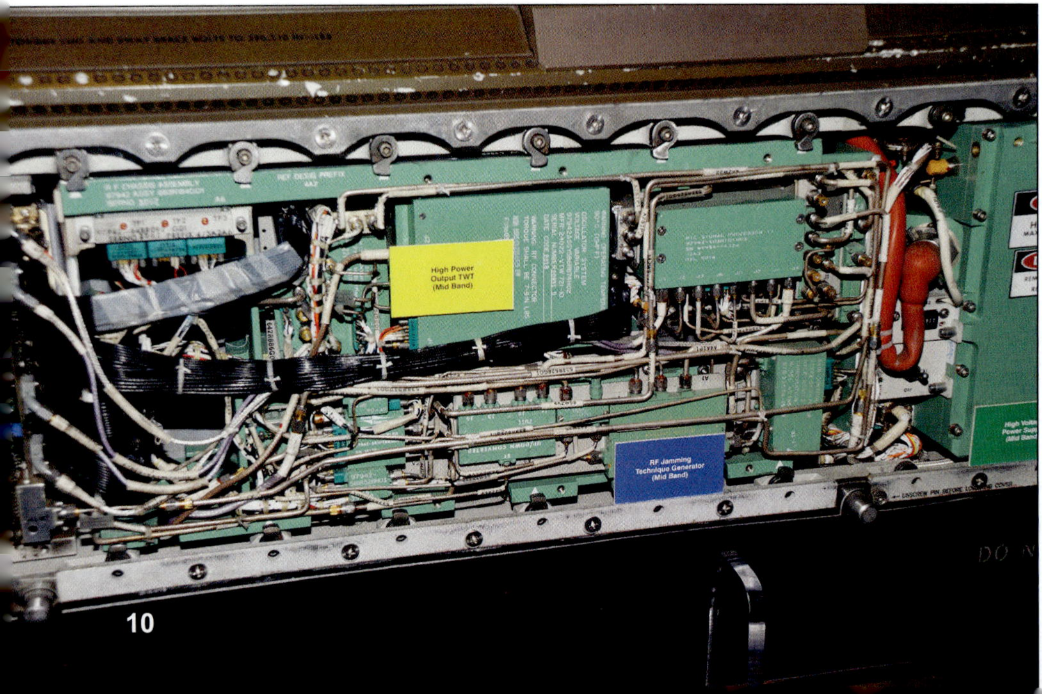

The rear section of the ECM pod consists of a pod chassis with two transmitter sections and slide-in circuit cards. There is a similar transmitter section located in the front of the pod. (John Gourley)

The Raytheon AN/ALQ-184 ECM jammer pod is mounted on the lower centerline of an F-16 Fighting Falcon aircraft. The pod is 635 pounds and 11 feet long. It is used on aircraft such as the A-10, F-15, F-16 and C-130. (John Gourley)

An F-16 Fighting Falcon from the South Dakota Air National Guard (ANG) takes off at Nellis Air Force Base (AFB) with AN/ALQ-184 ECM pod mounted on the lower fuselage centerline. The pod had major improvements from previous designs. (Jamie Hunter)

Airman 1st Class Robert Wulff with the 354th Maintenance Squadron troubleshoots a high voltage power supply on an AN/ALQ-184 ECM pod at Eielson Air Force Base in Alaska on 15 November 2006. (Staff Sgt. Josua Strang / U.S. Air Force)

Two A-10 Thunderbolt II "Warthog" aircraft from the Idaho ANG fly with AN/ALQ-184 ECM pods mounted on Station 1 wing pylons. The AN/ALQ-184 had 11 versions and is the primary jamming pod used by Air Force. (Master Sgt. Becky Vanshur / Idaho ANG)

The EC-130H Compass Call aircraft is primarily an Airborne Early Warning (AEW) aircraft that was expanded to include ECM radar jamming capabilities to work with other aircraft. Fourteen C-130 aircraft were modified to Compass Call planes. (Maj. Dale Greer / USAF)

An EC-130H Compass Call aircraft is prepared for a night mission in support of Operation Enduring Freedom at Bagram Air Base in Afghanistan. EC-103H are currently based at Davis-Monthan AFB in Arizona. (Master Sgt. Lance Cheung / U.S. Air Force)

EC-130H Compass Call

In recent years, multi-mission aircraft were developed to handle ECM functions as one of the different mission tasks. One such aircraft is a variation of the Lockheed C-130 Hercules that has been significantly modified by the Air Force to perform a number of electronic warfare tasks such as Airborne Early Warning (AEW). It can also handle some aspects of ECM radar jamming capabilities.

The Compass Call had it first flight in 1981 and was delivered to the Air Force in 1982. It reached its initial operations capability in 1983. The aircraft has been deployed to a number of war theaters from its introduction to now. It is currently used in Iraq and Afghanistan.

The EC-130H aircraft is manned by 13 crew members. Four members handle flight and navigation while nine other crew members operate the electronic attack mission equipment that is installed permanently in the cargo area of the aircraft. The mission crew includes the electronic warfare office who is also the mission crew commander, an electronic warfare officer, a mission crew supervisor, four analysis operators, one acquisition operator, and an airborne maintenance technician.

Compass Call aircraft are assigned to the Air Combat Command with the primary base at the 55th Electronic Combat Group (ECG) from Davis-Monthan Air Force Base in Arizona. All 14 Compass Call aircraft remain in service.

Staff Sgt. Matt Auble, a crew chief with the 43rd Expeditionary Electronic Combat Squadron (EECS), marshals an EC-130H Hercules into position after a mission in May of 2003. The EC-130H Compass Call aircraft saw significant war action during Operation Iraqi Freedom. (Airman 1st Class Kevin A. Camara / U.S. Air Force)

EA-18G Growler and Beyond

At the beginning of the 21st century, the U.S. Navy started to replace the EA-6B as the primary ECM platform. In 2001, Boeing successfully completed an initial flight demonstration of an F/A-18F Super Hornet fitted with the AN/ALQ-99 electronic-warfare system to serve as the EA-18 ECM concept aircraft.

In December of 2003, the Navy awarded a development contract for the EA-18G. The EA-18G Growler was designed by Boeing and used the F-18 Super Hornet as the baseline aircraft. In contrast to the EA-6B Prowler, the Grower can fly about 500 mph faster and reach 1,915 km/h (about Mach 1.8).

Production of the EA-18G Growler began in 2007. The aircraft entered operational service in late 2009. The Growler's flight performance is similar to the F/A-18E/F Super Hornet with the same engine.

Because the Growler was born from a fighter airframe, it performs escort jamming for other Navy fighter aircraft such as the F/A-18. The only active U.S. aircraft that could outpace a Growler are F-15, F-16, F-22 and F-35.

EA-18G aircraft typically carry three jamming pods – one low band and two high band. The pods weigh 950 pounds each. Each pod carries two 2-kilowatt transmitters, a control computer and various electronic modules such as power supplies, as well as exciters and drivers.

The EA-18G was first used in combat during Operation Odyssey Dawn from 19 to 31 March 2011 to enforce the United Nations (UN) no-fly zone over Libya. A total of five EA-18Gs of Electronic Attack Squadron 132 (VAQ-132) were redeployed from Italy to support Libya operations and Iraqi operations (Operation New Dawn).

In addition to the 150 EA-18G aircraft ordered by the U.S. Navy, the Royal Australian Air Force (RAAF) ordered a dozen EA-18G aircraft as their primary ECM aircraft. The RAAF's initial EA-18G first flew in July 2015. These were the only U.S. ECM aircraft sold to a foreign country.

The AN/ALQ-99 system was used for more than 30 years. Obsolete electrical components are difficult to replace as electronic technology improves and older technologies are discontinued. Consequently, the Next Generation Jammer (NGJ) was proposed by the Navy to replace the current AN/ALQ-99 system.

In July 2013, the Navy awarded the NGJ contract to Raytheon over Northrop Grumman, the developer of the current AN/ALQ-99 system. For the first phase, Raytheon will deliver 15 engineering development units as well as 14 pod structures for airworthiness certification. The NGJ will use a smaller antenna that is divided into quadrants for all-around coverage while it retains the capability of highly direction jamming. Initial operational capability (IOC) is planned for early 2021.

With continuous radar technology improvements, ECM equipment must improve and upgrade.

The U.S. Air Force has not pursued dedicated ECM aircraft since the EF-111A Raven was retired. Instead, emphasis was placed on stealth aircraft. However, the Navy continues to develop dedicated ECM aircraft as an integral part to support the carrier air wing.

An EA-18G Growler prepares for a return flight to NAS Whidbey Island after a one-day stopover in New York after a deployment in Libya in November of 2011. The Growler is the heir apparent for future ECM missions. The aircraft is based on the F-18 Super Hornet aircraft, and it has already been deployed into combat by the Navy. (Ken Neubeck)

An AN/ALQ-99 jamming pod is attached to the wing pylon of an EA-6B on display at an airshow at Patrick Air Force Base in Florida. AN/ALQ-99 jamming pods, such as this high band jamming pod, have been used for ECM aircraft such as the EA-6B and EA-18G. It has been in service for more than 30 years. (John Gourley)

An EA-6A from the VAQ-33 "Firebirds" flies as one of four EA-6A aircraft assigned to this squadron in 1978. Modified from existing A-6 Intruder airframes, the EA-6A first flew for the U.S. Navy in 1963. In addition to the Navy, the Marines also flew the aircraft, which served in Vietnam. (J.W. Hawkins collection)

An EA-6B prepares to be towed as a static display at the American Airpower Museum in New York for an airshow exhibit in 2012. The EA-6B first flew in Navy service in 1971 and saw action over Vietnam as well. As the jammer missions increased in complexity, the crew size increased from two in the EA-6A to four in the EA-6B. (Ken Neubeck)

EA-6A Intruder

Wingspan	53 feet
Length	55 feet, 6 inches
Height	14 feet, 6 inches
Empty Weight	27,769 pounds
Combat Weight	54,571 pounds
Powerplant	Two Pratt & Whitney J52-P-6A turbojet engines
ECM system	AN/ALQ-86
Maximum Speed	631 miles per hour
Service Ceiling	37,800 feet
Range	2,012 miles
Crew	2
Number of aircraft built	28

EA-6B Prowler

Wingspan	53 feet
Length	59 feet
Height	16 feet, 8 inches
Empty Weight	31,160 pounds
Combat Weight	61,500 pounds
Powerplant	Two Pratt & Whitney J52-P-408A turbojet engines
ECM system	AN/ALQ-99
Maximum Speed	651 miles per hour
Service Ceiling	37,600 feet
Range	2,922 miles
Crew	4
Number of aircraft built	170

An F-105G from the 388th TFW flies a mission over Vietnam. Beginning in 1970, F-105F Thunderchiefs were modified into F-105G. ECM pod sections were added to the fuselage, while radar homing and warning equipment were installed. (National Archives)

F-105F/G Thunderchief (Wild Weasel III / IV)

Wingspan	34 feet, 11 inches
Length	67 feet
Height	20 feet, 1 inch
Empty Weight	30,419 pounds (F) / 31,279 pounds (G)
Combat Weight	54,580 pounds
Powerplant	One Pratt & Whitney J75-P-19W turbojet engine
ECM system	AN/ALQ-71 (F) / AN/ALQ-105 (G)
Maximum Speed	1,372 miles per hour
Service Ceiling	48,500 feet
Range	1,494 miles
Crew	2
Number of aircraft modified	86 (F) / 63 (G)

An F-4G Phantom II, Wild Weasel V, banks during a test flight in 1981 with an ECM pod mounted on the fuselage and four different AGM missiles situated under the wing pylons. The F-4G assumed the Wild Weasel role after the F-105G was retired. (USAF)

F-4G Phantom II (Wild Weasel V)

Wingspan	38 feet, 4.5 inches
Length	63 feet
Height	16 feet, 6 inches
Empty Weight	30,328 pounds
Combat Weight	41,500 pounds
Powerplant	Two General Electric J-79-GE-17A turbojet engines
ECM system	AN/ALQ-173
Maximum Speed	1,472 MPH
Service Ceiling	60,000 feet
Range	1,494 miles
Crew	1,615 miles
Number of aircraft built	134

An EF-111A flies to Eglin AFB during Exercise Solid Shield in 1987. The EF-111A was modified from existing F-111A Aardvark aircraft and first flew in 1981. It served in Desert Storm in 1991 and other Eastern Europe engagements until it was retired in 1998. (USAF)

An EA-18G from VAQ-132 parks at New York's Republic Airport in November of 2011. The EA-18G is based on the F-18 Super Hornet design. It first flew in 2007. This aircraft saw action in Iraq and Afghanistan soon after it came into service. (Ken Neubeck)

EF-111A Raven

Wingspan	63 feet (spread), 32 feet (swept)
Length	76 feet
Height	20 feet
Empty Weight	55,275 pounds
Combat Weight	80,000 pounds
Powerplant	Two Pratt & Whitney TF3-P-3 (initial), TF3-P-9 (later) turbofan engines with afterburner
ECM system	AN/ALQ-99E
Maximum Speed	1,460 miles per hour
Service Ceiling	45,000 feet
Range	1,740 miles
Crew	2
Number of aircraft built	44

EA-18G Growler

Wingspan	44 feet, 8 inches
Length	60 feet
Height	16 feet
Empty Weight	33,094 pounds
Combat Weight	66,000 pounds
Powerplant	Two General Electric F414-GE-400 turbofan engines
ECM system	AN/ALQ-99
Maximum Speed	1,190 miles per hour
Service Ceiling	50,000+ feet
Range	1,275 miles
Crew	2
Number of aircraft built	100+ (still in production)

EA-6A Intruder

An EA-6A from the VMAQ-2 "Playboys" of Cherry Point Marine Corps Aviation Reconnaissance Association (MCARA) flies over a river delta in North Carolina in 1979. The EA-6A Intruder represents a major ECM mission change. The aircraft was designed and constructed for the ECM mission rather than modified from an existing aircraft originally designed for other missions. After the initial lot of modified A-6A airframes became EA-6A Intruders, there was a production run of the purpose-built aircraft. The EA-6A features a two-man crew, which consists of a pilot and an ECM operator. (Grumman archives via Northrop Grumman History Center)

A pair of A-6E Intruders with Attack Squadron 65 (VA-65) "Tigers" fly in 1987. The A-6 Intruder was first introduced in 1960 and became the basic airframe for the EA-6. (Navy)

An EA-6A is serviced during a stop at Patrick AFB in 1990. The EA-6A jammer equipment was incorporated into the aircraft and featured external changes. (John Gourley)

The tail section of an A-6 Intruder display features the original tail configuration. The tail was later modified significantly to house the radar antenna receiver. (Ken Neubeck)

An EA-6A is chocked on a hardstand. The modified tail section holds the radar antenna, which is the most significant external change from the A-6A airframe. (John Gourley)

Radar equipment fills the EA-6A's nose. The attack radar antenna for the AN/APQ-112 radar system is the dish with a "NO HAND HOLD" warning. The antenna assembly and gimbal assembly is the AN/APQ-92 system's search radar antenna. Behind the antennas are the electronic boxes for these radar systems. (Northrop Grumman History Center)

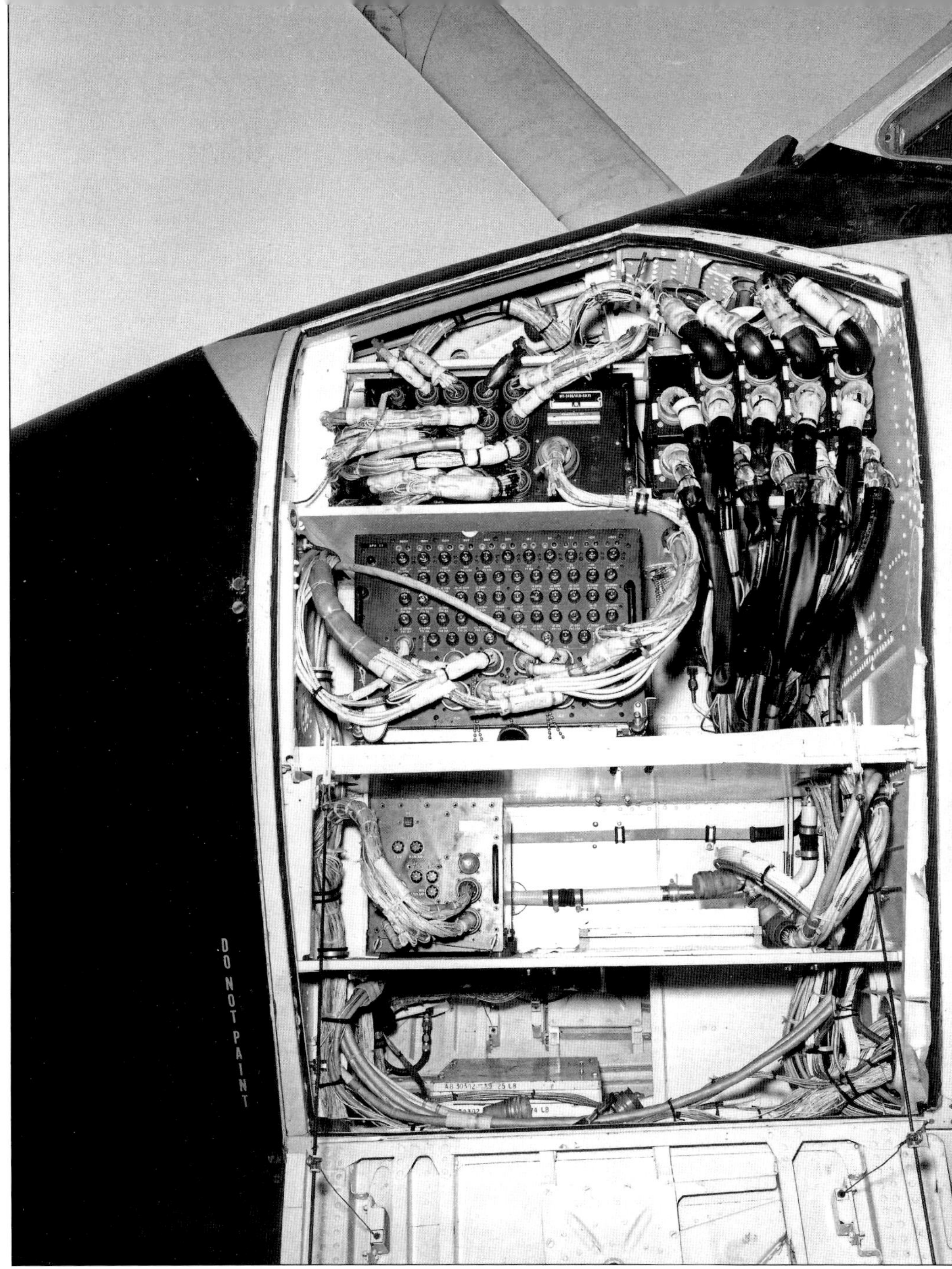

Additional electronic equipment – which include a circuit breaker panel, junction boxes for wire harnesses, and cable assemblies – are packed into the left side of the fuselage, below the canopy section. The access panel to this area is hinged at the bottom and allows for quick release for ground maintenance. (Northrop Grumman History Center)

An EA-6A Intruder cockpit has more space prior to the installation of ejection seats. The left side is the pilot station has standard flight control equipment and fuel gages. The ECM operator's station on the right includes scopes to track radar signals along with various control knobs to shape and control jammer signals for the AN/ALQ-86 ECM system used in the EA-6A. (Grumman Archives via Northrop Grumman History Center)

An EA-6A prototype aircraft parks on the USS *Kitty Hawk* deck with its wings folded and jammer pods attached. (Grumman archives via Northrop Grumman History Center)

Carrying two types of jammer pods, an EA-6A prototype catapults from the USS *Kitty Hawk*. (Grumman archives via Northrop Grumman History Center)

An EA-6A prototype deploys its arresting hook to land on the USS *Kitty Hawk*. (Grumman archives via Northrop Grumman History Center)

An original EA-6A is parked near the catapult in 1963. Lines are marked on the fuselage as photographic references. (Grumman archives via Northrop Grumman History Center)

EA-6A 148618, one of two EA-6A prototypes, moves into position on the carrier flight deck behind a Grumman C-2 Greyhound aircraft in 1965. The prototype actively served the Navy until 1980 when it was put into storage at the Aerospace Maintenance and Regeneration Center (AMARC - also known as the 309th Aerospace Maintenance and Regeneration Group or "The Boneyard") in Arizona. It returned to service in 1981 and served until 1993. It is presently preserved at Naval Air Station (NAS) Key West in Florida. (Grumman Archives via Northrop Grumman History Center)

One of two Pratt & Whitney J52-P-6A engines that power the EA-6A receives maintenance at Grumman in New York. (Grumman archives via Northrop Grumman History Center)

Navy crewmen attach a jamming pod to an EA-6A assigned to the VAQ-209 "Star Warriors" squadron. (Grumman archives via Northrop Grumman History Center)

An early jamming pod is mounted on an EA-6A centerline. The pod was later upgraded to a ram-air design. (Grumman archives via Northrop Grumman History Center)

The original centerline jamming pod does not have ram air cooling for the electronics. (Grumman archives via Northrop Grumman History Center)

An EA-6A Intruder is flanked by two A-6A aircraft in production at Grumman in New York. (Grumman archives via Northrop Grumman History Center)

Intruder EA-6A 156984 of the Electronic Attack Squadron 209 (VAQ-209) "Star Warriors" undergoes ground maintenance while on the tarmac of Patrick Air Force Base in February of 1990. (John Gourley)

An EA-6A Intruder flies in December 1978. The Marine Tactical Electronic Warfare Squadron 2 (VMAQ-2) "Playboys" use the CY tail code and is based at Marine Corps Air Station (MCAS) in Cherry Point, North Carolina. (Staff Sgt. Jomp / USMC)

An EA-6A Intruder shares the deck of aircraft carrier USS *Ranger* with two EA-6B Prowlers in June of 1980. (Ron Wood / U.S. Navy)

The most visible external difference between the A-6A and the EA-6A is its enlarged tail section with a black trim. (Northrop Grumman History Center)

An EA-6A from the VAQ-209 "Star Warriors" gets auxiliary power provided by ground service carts at NAS Whidbey Island in Washington. (Northrop Grumman History Center)

Intruder EA-6A 156984 from the reserve squadron VAQ-209 "Star Warriors" rests in a hanger in NAS Norfolk, Virginia, in April of 1979. (Northrop Grumman History Center)

An EA-6A from the VAQ-309 "Axemen" prepares for a catapult launch with a low-band jamming pod from the USS *Enterprise* in August of 1989. (Richard L. Larson / Navy)

AEA-6A 156979 performs a test flight over eastern New York in 1963. It was lost after a combat operation over North Vietnam in April of 1972. Both the pilot and ECM operator were killed in action. The EA-6A ECM-version Intruder was modified from the A-6A airframe and first flew in 1963. Grumman built a total of 28 EA-6A aircraft, which replaced the U.S. Marines' F3D-2Q Skyknight and saw action over Vietnam. The last EA-6A in Marine service was retired in 1993. (Grumman archives via Northrop Grumman History Center)

EA-6B Prowler

EA-6B 158817 from the VAQ-137 "World Famous Rooks" squadron of NAS Oceana flies during a mission. The bird insignia of the VAQ-137 squadron is emblazoned on the tail of the aircraft. The aircraft went down in the Pacific Ocean off of the coast of California while it conducted a night training mission from the aircraft carrier USS *Ranger* in September of 1978. The pilot, Lt. Cmdr. Clark Bruce and ECMO 2, Lt. John R. Babione, both died on impact while ECMO 1, Lt. Cmdr. W. L Waterman, ejected successfully on impact. The EA-6B Prowler was the primary U.S. ECM aircraft during the Persian Gulf War in 1991. During the war, 39 EA-6B aircraft flew 1,650 sorties and 4,600 flight hours with no losses. (Northrop Grumman History Center)

An EA-6B rests at Republic Airport in New York. It has special gold-plated canopy glass for sun protection. Windshield sections were fitted with non-tinted glass. (Ken Neubeck)

The forward and aft cockpit canopy sections are in the open position during a stop in NAS Jacksonville, Florida, in 1998. The canopies are hinged at the back. (John Gourley)

The ECM Operator 1 crew member in the front right position prepares for takeoff. Crew boarding steps are built into the front of the engine inlet section. (Ken Neubeck)

The aft cockpit section contains the ECM Operator 2 crew member on the left and the ECM operator 3 crew member in the right position. (Ken Neubeck)

The EA-6B has four cockpit stations in 1977. The pilot sits in the left front position and the three electronic countermeasures officers (ECMO) operators work in the other positions. Flight control gages are located in the left side only. This particular cockpit is the forward EA-6B cockpit with Improved Capability (ICAP), which superseded the EXCAP version. As in the EXCAP version, the ICAP version included eight bands of coverage, up to 10 GHz range. (Grumman archives via Northrop Grumman History Center)

The aft cockpit for an ICAP-equipped E-A-6B is ready for a mission. The ICAP aircraft improved the workload for the ECMO 1 position in the forward cockpit by moving the tactical jamming system functions from the ECMO 1 operator to the ECMO 2 and ECMO 3 operators. There are no flight controls for the aft cockpit. The large panel on the top of the console is the main circuit breaker panel. (Grumman archives via Northrop Grumman History Center)

The ECM operator No. 1 position is in the right front station. The main display has a hood attachment to prevent outside glare. (John Gourley)

A tinted canopy protects the rear stations and large circuit breaker panel. (Ken Neubeck)

The ECM operator No. 3 position is in the right rear station. The circuit breaker panel for all of the avionics is located on top of the console. (John Gourley)

Lt. j.g. Christopher Rowan (left) and Lt. Cmdr. Matt Shipley (right) of VAQ-141 conduct pre-flight checks on the AN/ALQ-99 Tactical Jamming System before launching from the USS *Theodore Roosevelt* in the Atlantic Ocean on 26 March 2005. (Eben Bootby / Navy)

ECM operator No. 3 will sit in the right rear position to operate the tactical jamming suite (TJS) while the EA-6B pilot strictly flies the aircraft. (John Gourley)

ECM operator No. 2 will sit in the left rear position and also operate the TJS in addition to the ground mapping AP-130 system. (John Gourley)

Rear Adm. J.R. Haley, Carrier Strike Group 5 commander, enters an EA-6B before piloting a flight with the VAQ-136 "Gauntlets" in the South China Sea. (William Pittman / Navy)

An EA-6B is parked at Republic Airport in 2012. The roles EA-6B crew members remained consistent throughout the history of the Prowler and various upgrades. (Ken Neubeck)

An EA-6B flies over clouds. It has a TR tail code to indicate that it is from the Replacement Air Group Squadron VAQ-129 "Vikings." (Northrop Grumman History Center)

An EA-6B with the Navy's Fleet Replenishment Squadron VAQ-129 "Vikings" banks to show a full complement of ECM jamming pods. (Northrop Grumman History Center)

A Marine EA-6B 161244 Prowler prepares for takeoff from Patrick Air Force Base in Florida in June of 1994. (John Gourley)

Four EA-6B Prowlers fly in formation. Three of the aircraft are from the VAQ-129 "Vikings" and are yet to be deployed to a Naval Air Station. (Northrop Grumman History Center)

Sailors Cameron Brown (left) and Eric Augustine (right) direct an EA-6B aboard aircraft carrier USS *John C. Stennis* on 6 January 2013. (Kenneth Abbate / U.S. Navy)

An EA-6B is parked on a hardstand. Because the EA-6B is a carrier-based aircraft, it must fold its wings at the wing centerline to park on the carrier deck. (Ken Neubeck)

An EA-6B aircraft from the VAQ-130 "Zappers" squadron launches from the USS *Harry S. Truman* (CVN 75) in the Atlantic Ocean with a complement of four high-band jamming pods located on the wing pylons in 2010. Three EA-6B aircraft with folded wings are parked on the side of the deck with similar high-band jamming pod configurations. This was the last deployment for the EA-6B with the "Zappers" Electronic Attack Squadron. (Kilho Park / U.S. Navy)

EA-6B 161285 from VAQ-129 parks on a hardstand at MacDill AFB in Florida in 1993. It is configured with no ECM pods or fuel pods on the wing stations. (John Gourley)

EA-6B 163401 from Air Development Squadron 5 (VX-5) is chocked in Sanford, Florida, with only the two high-band ECM pods mounted on outboard pylons. (John Gourley)

EA-6B 163049 is displayed at Cecil Field in 1993. It is equipped with no ECM pods, but carries three fuel pods under the wing stations and fuselage station. (John Gourley)

EA-6B 163259 visits Daytona, Florida, in 1997. It was equipped with two high-band ECM pods on outboard pylons and a fuel pod on the fuselage station. (John Gourley)

An EA-6B is parked at Sanford AFB in 1995. It carries an AGM-88 High Speed Anti-Radiation Missile (HARM) air-to-surface missile on its outboard pylon . (John Gourley)

An EA-6B carries a trainer version of an AGM-88 HARM air-to-surface missile. The combat version of the missile is used to attack verified radar sites. (John Gourley)

Airman Francisco Escobedo assigned to the (VAQ-136) "Gauntlets" prepares a AGM-88 HARM on an EA-6B Prowler aboard the aircraft carrier USS *Kitty Hawk* (CV 63) in support of Operation Iraqi Freedom on 28 March 2003. (Todd Frantom / U.S. Navy)

U.S. Navy students conduct exercises to remove a captive air training missile 88 (CATM-88) from an EA-6B Prowler at the Center for Naval Aviation Technical Training Unit at NAS Whidbey Island in Washington on 19 April 2010. (Nardelito Gervacio / U.S. Navy)

An EA-6B Prowler with the VAQ-140 "Patriots" refuels in flight from a specially-fitted F/A-18E Super Hornet of the Strike Fighter Squadron (VFA) 143 "Pukin Dogs" from USS *Dwight D. Eisenhower* on 19 October 2012. (Nathanael Miller / U.S. Navy)

Two EA-6Bs from the VAQ-136 "Gauntlets" make a final deployment flight over Naval Air Facility (NAF) Atsugi in Ayase, Japan, on 14 February 2012. (Justin Smelley / U.S. Navy)

An EA-6B Prowler equipped with three jamming pods and two external fuel tanks flies near the USS *Enterprise* on 21 September 2012. (Scott Pittman / U.S. Navy)

Two EA-6B Prowlers from the VAQ-134 "Garudas" deploy arresting hooks before they land on the aircraft carrier USS *Carl Vinson* (CVN 70) as F/A-18 Hornets circle overhead in the Arabian Gulf on 27 February 2012. (Andrew K. Haller / U.S. Navy)

EA-6B 162938 rests on the runway at Republic Airport in New York in the Spring of 2012. This aircraft was last assigned to the Whidbey Island squadron and was designated for retirement. It was decommissioned, and it is currently on display at the American Airpower Museum at Republic Airport in New York. It will eventually be featured as a gate guard display at the Northrop Grumman facility in Bethpage in New York. (Ken Neubeck)

Grumman began production of the EA-6B in New York in 1966 and the last aircraft rolled out in 1991. (Ken Neubeck)

As of 2016, there are fewer than 50 EA-6B Prowler aircraft in service. Final retirement of the EA-6B planned for 2019. (Ken Neubeck)

The AN/ALQ-99 high band pod is located below and to the inside of the wing fold mechanism of the EA-6B Prowler. (Ken Neubeck)

Cmdr. Michael MacNicholl of VAQ-137 inspects his EA-6B Prowler on the USS *Enterprise* on 15 October 2012. The rear of the AN/ALQ-99 pod is open. (Scott Pittman / U.S. Navy)

EA-6B 159908 is configured for the high band mode only as it contains the two high band pods mounted on the outside wing pylons. (Ken Neubeck)

An EA-6B approaches the USS *Abraham Lincoln*. It is configured for all modes because it carries two high-band pylon pods and a low-band fuselage pod. (Brian Morales / Navy)

The interior walls of the AN/ALQ-99 jammer pods are made of woven fiberglass material to send and receive radio waves during operation. The antenna is a horn-style, high-band antenna. It can be steered in different directions. Four distinct channels line the inside wall of the antenna to tune antenna frequencies. A similar antenna is located in the front portion of the pod. A low-band jamming pod is mounted under the fuselage and features similar internal construction. (John Gourley)

Each AN/ALQ-99 jammer pod generates its own power through a ram air turbine, which consists of a small propeller that drives an electric generator. (Ken Neubeck)

Holes behind the ram-air-turbine generator allow a stream of air to provide ram air cooling to flow through the pod and remove hot air generated by the electronics. Hot air vents out through the open rear radome of the pod. (Ken Neubeck)

The protrusion at the base of the aerial refueling pod is the ALQ-126 receiving antenna used for the deception electronic countermeasures (DECM). (Ken Neubeck)

The VHF (very high frequency) antenna and the IFF (identification, friend or foe) antenna are located in front of the nose landing gear on the lower fuselage. (Ken Neubeck)

The EA-6B fin pod contains countermeasure receivers located toward the front of the pod and receiving antennas located toward the rear of the pod. (John Gourley)

A maintenance crew performs Quality Assurance checks on an EA-6B Prowler from VAQ-129 at NAS Whidbey Island in Washington on 5 April 2005. (Chris Perkins / Navy)

EA-6B 162938 has the Viking motif on the rudder and tail code marking NJ, which is the VAQ-129 Squadron based at NAS Whidbey Island in Washington. (Ken Neubeck)

EA-6B 160437 has the wolf motif on the tail tip and tail code marking MD, which is the VMAQ-3 Squadron based at MCAS Cherry Point. (John Gourley)

EA-6B 161349 has the seahawk motif on tail tip and tail code marking RM, which is from the VMAQ-4 Squadron based at MCAS Cherry Point. (John Gourley)

EA-6B 163034 has the scorpion motif on the rudder and is from the VAQ-132 Squadron. (John Gourley)

The large Shadowhawk motif is among multiple markings on the tail of EA-6B 163046 from NAS Whidbey Island in Washington on 12 August 2012. (John Gourley)

An EA-6B bears various markings on the lower fuselage that signify completed ECM missions and the destruction of one radar site. (John Gourley)

Nicknamed the "beer can," the AN/ALQ-126 receiving antenna is the cylindrical structure that protrudes from the bottom of the fin pod radome, directly under the number 4. The AN/ALQ-126 Deception Electronic Countermeasures (DECM) system is standard for the Improved Capability (ICAP) version EA-6B Prowlers and is used to delay and regenerate radar signals form enemy tracking radars. The system is carried by several Navy aircraft in addition to the EA-6B and can deliver more than 1 kilowatt of jamming power per band in the pulse mode. (John Gourley)

Crewmen prepare an EA-6B Prowler for catapult-assisted takeoff from the carrier USS *Abraham Lincoln.* (Del Laughery)

An EA-6B crew studies paperwork for a mission task and prepares for takeoff. (Del Laughery)

Two EA-6B Prowlers from the NAS Whidbey Island are loaded with crew and prepared for takeoff from the carrier USS *Abraham Lincoln.* (Del Laughery)

The arresting hook of an EA-6B Prowler is about to engage the arresting cable on the deck of USS *Abraham Lincoln*. (Del Laughery)

An EA-6B of the VAQ-137 "Rooks" launches from the USS *Theodore Roosevelt* deck to support Operation Enduring Freedom on 20 January 2002. (Amy DelaTorres / U.S. Navy)

An EA-6B of the VAQ-138 "Yellow Jackets" is parked on the deck of the USS *John C. Stennis* during Operation Enduring Freedom on 27 February 2002. (Joshua Word / Navy)

An EA-6B is about to engage the arresting cable on the USS *John C. Stennis* as it lands after mission in January 2002. (Quinton Jackson / U.S. Navy)

An EA-6B from VAQ-138 joins an F-18 Hornet from the Marine Fighter Attack Squadron 314 (VMFA-314) onboard the *Stennis* in December of 2001. (Troy M. Latham / U.S. Navy)

An EA-6B Prowler from Naval Air Station Whidbey Island lands on the USS *Abraham Lincoln* during field exercises in 2008. (Del Laughery)

An EA-6B from the VAQ-135 "Black Raven" squadron at NAS Whidbey Island extends its arresting hook as it lands on the USS Nimitz. (John Phillip Wagner Jr. / U.S. Navy)

A retired EA-6B test aircraft is propped into place on the radar test range at NAS Patuxent River in Maryland. The pivot base is stored on top of the aircraft. (Ken Neubeck)

An EA-6B aircraft flies over Colorado in May of 2011. The aircraft is configured with two high-band pods under the wings. (Ben Peck)

F-105F Thunderchief (Wild Weasel III)

F-105F 62-4436 prepares for a mission. It would eventually be converted to a F-105G configuration. The aircraft was lost during a SAM-suppression mission over North Vietnam in November 1970, however both the pilot and ECM operator were rescued. The development of the Thunderchief F-105F, Wild Weasel III, aircraft to fly suppression missions against SAM radar sites in Vietnam is a key development in the ECM mission. Initially assigned to F-100F aircraft, the F-105F two-seat aircraft was modified to become a SAM hunter by carrying a ECM jammer pod on the outboard right wing pylon station to jam North Vietnamese radar sites. The aircraft would attack these sites with AGM-45 Shrike missile on the left outboard wing pylon station. (Fairchild Republic Archives via Cradle of Aviation Museum)

A new F-105F aircraft rolls out of the Fairchild Republic facility in 1965. The aluminum finish was later painted in Vietnam. (Fairchild Republic archives via Cradle of Aviation)

Early F-105F ECM aircraft were armed the AGM-45 Shrike missile and carried bombs in the internal bay. (Fairchild Republic archives via Cradle of Aviation Museum)

A Thunderchief F-105F is outfitted with ECM pod on the outboard weapon pylon of the right wing in Vietnam. (Fairchild Republic archives via Cradle of Aviation)

A F-105F, Wild Weasel III, is equipped with one AGM-45 missile on the outside pylon to attack SAM radar sites. (Fairchild Republic archives via Cradle of Aviation Museum)

The rear cockpit of a F-105F includes basic flight instruments for the rear crew member to fly the aircraft. Additional equipment was added to the rear cockpit position for ECM work. (Fairchild Republic archives via Cradle of Aviation Museum)

The F-105G rear cockpit station is modified for ECM operators. Additional equipment includes an APR-35 display and controls in the center. (Ben Knowles via Dennis R. Jenkins collection)

The crew stations consist of a pilot in front and an ECM operator in the back section. The ECM operator also has flight controls to pilot the aircraft. (Ken Neubeck)

F-105G Thunderchief (Wild Weasel IV)

Prototype F-105G number 2, serial number 62-4434, flies with the new ECM setup devised by Fairchild Republic, Westinghouse, and the U.S. Air Force. The ECM jammer pod was split in half and moved to the F-105G fuselage. This change freed a wing pylon for another weapon to attack SAM radar sites in North Vietnam. The new system was originally designated Quick Reaction Capabilities 380 (QRC-380) before its redesignation as the AN/ALQ-105. There was no EF version of the F-105G. (National Archives)

The ECM pod mounted on the lower fuselage of the F-105G has access doors to facilitate work on the pod. (Fairchild Republic archives via Cradle of Aviation Museum)

Half of an ECM pod is mounted on each side of the F-105G aircraft on the lower fuselage below the engine intake section. (Ken Neubeck)

Westinghouse split the AN/ALQ-101 pod in half to create two separate AN/ALQ-105 pods to mount on the fuselage. (Fairchild Republic archives via Cradle of Aviation Museum)

The black composite material on the ECM pod allows the receiver to detect radar signals while it protects delicate electronics from high-speed wind exposure. (Ken Neubeck)

The ECM pod incorporated an indention to allow the landing gear doors to function properly. (Fairchild Republic archives via Cradle of Aviation Museum)

Two ECM features associated with the F-105G are the ECM fuselage pod and an extended tail for the RHAW system. (Fairchild Republic archives via the Cradle of Aviation)

Additional components of the AN/APR-35 Radar Homing and Warning (RHAW) system include three circular antennas mounted on the front lower fuselage. (Ken Neubeck)

The top portion of the tail section was increased five inches to accommodate fairings for the Radar Homing and Warning (RHAW) system. (Ken Neubeck)

In addition to more space on one F-105G wing pylon, the body-mounted AN/ALQ-120 ECM pod facilitated better performance than the wing-mounted ECM pod. The F-105G was in Vietnam until 1974, then served with the Georgia ANG until 1983. (Ken Neubeck)

The F-105F and F-105G aircraft nose portion consisted of the R-14 radar, which included a dish antenna with the electronics located behind it. The nose radome was made of non-metallic material to allow radar signals to pass through the structure. (USAF)

Thunderchief F-105G 66-4432 rolls out of Fairchild Republic's plant in New York in February 1974. (Fairchild Republic archives via Cradle of Aviation Museum)

An F-105G with a WW tail code taxis down the runway at Korat air base in Thailand with ordnance on three pylons and fuel cell on the remaining pylon. (National Archives)

An F-105G flies while configured with fuel cells on the inboard pylons, training missiles on the outboard pylons and training bombs on the centerline. (Sheryl D. Barnett / USAF)

After Vietnam service, remaining F-105G aircraft were assigned to the 128th TFS of the 116th Tactical Fighter Wing in Georgia during the 1980s. (Sheryl D. Barnett / USAF)

F-4G Phantom II (Wild Weasel V)

When the F-105 ended production in 1964 and losses drained the F-105 inventory over Vietnam, the U.S. Air Force needed an ECM aircraft replacement. The F-4G model filled this role at the speed of Mach 2.2. A total of 134 F-4G models were converted from the F-4E Phantom II. The first F-4G flew in 1975 while actual service began in 1978.

F-4Gs were deployed to three active wings. One wing was stationed at George AFB in Victorville, California; one wing was assigned to USAFE (U.S. Air Forces in Europe) at Spangdahlem Air Base in Germany; and the other wing went to PACAF (Pacific Air Forces) at Clark AFB in the Philippines. There was no EF version of the F-4G.

F-4Gs from both George AFB and Spangdahlem Air Base saw combat during Operation Desert Storm in 1991. These planes successfully protected strike and attack aircraft from enemy air defenses. During the conflict the F-4G saw heavy use with only a single loss. The only F-4G loss was from Spangdahlem Air Base and crashed in Saudi Arabia while it returned from a mission. However, both the pilot and EWO safely ejected.

After Desert Storm, some George Air Force Base aircraft were assigned to the 124th Wing of the Air National Guard at Boise, Idaho. Aircraft from Spangdahlem, Clark, and the remainder from George were assigned to the 57th Fighter Wing (active Air Force) assigned to Nellis AFB at Las Vegas. The aircraft remained in service until 1996 when both squadrons participated in frequent deployments to Saudi Arabia and Turkey in support of various missions over Iraq where no-fly zones were enforced. By this time, the F-4G was the last operational variant of the Phantom II in U.S. forces. Eventually, a number of the F-4G airframes would be used as target drones.

An F-4G from Spangdahlem Air Base flies over Germany in 1990. The F-4G Phantom, Wild Weasel V, aircraft was developed in 1975 after significant losses of the F-105G Thunderchief, Wild Weasel IV, aircraft over Vietnam. (Staff Sgt. Fernando Serna / USAF)

The tail section of the F-4G was modified to include an APR-47 high-band tailfin antenna. (John Gourley)

F-4G 69-0263 uses an AN/ALQ-184 ECM pod that is mounted to the lower fuselage of the aircraft. (John Gourley)

Two 35th Tactical Fighter Wing F-4G Phantom II aircraft pass over the Saudi desert while on a training flight during Operation Desert Shield. The aircraft carry AGM-88 HARM high-speed anti-radiation missiles on their inboard wing pylons. (U.S. Air Force)

An F-4G from the 563rd TFS banks with an AN/ALQ-119 ECM pod attached to the left side pylon and a AGM-78 anti-radiation missile located on the inboard left wing pylon. It carries an auxiliary fuel tank on the centerline. (Master Sgt. Paul Harrington / USAF)

Ground crews service an F-4G Phantom II, Wild Weasel V, from the 81st TFS on the flight line beside an F-16 Fighting Falcon before it departs for Isa Air Base in Bahrain during Operation Desert Shield in late 1990. (Tech. Sgt. Fernando Serna / U.S. Air Force)

F-4G Phantom II 69-0248 from the 480th TFS (bottom) and an F-16C Fighting Falcon of the 52nd TFW (top) fly over the German countryside in June of 1989. The F-4G carries AGM-88 HARM and AIM-7 Sparrow missiles. (Master Sgt. Bill Thompson / USAF)

EF-111A Raven

EF-111A 66-0013 flies during a mission high above the clouds. It was the longest-serving F-111A Aardvark with the oldest serial number to be converted to the ECM configuration. The aircraft was officially designated as the Raven. It served with the 429th ECS during its service until it was retired in 1998. Deliveries of Mach 2.3 EF-111A aircraft began in 1981 and continued until 1985. Grumman converted a total of 42 F-111A aircraft into EF-111A models. The EF-111A participated in Operation Desert Storm in 1991 and supported U.S. operations in the Balkans in 1995 until it was finally retired in 1998. (Grumman Archives via Northrop Grumman History Center)

Aardvark F-111A 67-061 from Mountain Home AFB in Idaho flies with a load of bombs and served in Vietnam. The F-111A airframe was the basis for the EF-111A Raven. (USAF)

The EF-111A was modified from existing F-111A aircraft with the addition of ECM antennas located on the fuselage top and bottom and an enlarged tail. (John Gourley)

An unmodified F-111A Aardvark empennage shows the early 1960s lines. The EF-111A tail section was significantly modified to house the AN/ALQ-99E receiver. (Ken Neubeck)

The modified tail section of the EF-111A Raven holds the radar antenna and is the most significant external change from the basic F-111A airframe. (John Gourley)

The EF-111A jamming pod is mounted directly to the lower fuselage rather than extending from the aircraft. (John Gourley)

The distinctive enlarged tail tip includes the antenna and receiver for the AN/ALQ-99E system used in the EF-111A aircraft. (John Gourley)

The bladed band-1 and band-2 ECM antenna is mounted mid-fuselage on the EF-111A to avoid existing structures and antennas to maximize radiation pattern. (John Gourley)

Mounted on top the EF-111A fuselage is the AN/ALR-62 Track While Scan (TWS) antenna for the AN/ALQ-137 ECM system. An ALR is a Radar Warning Receiver. (John Gourley)

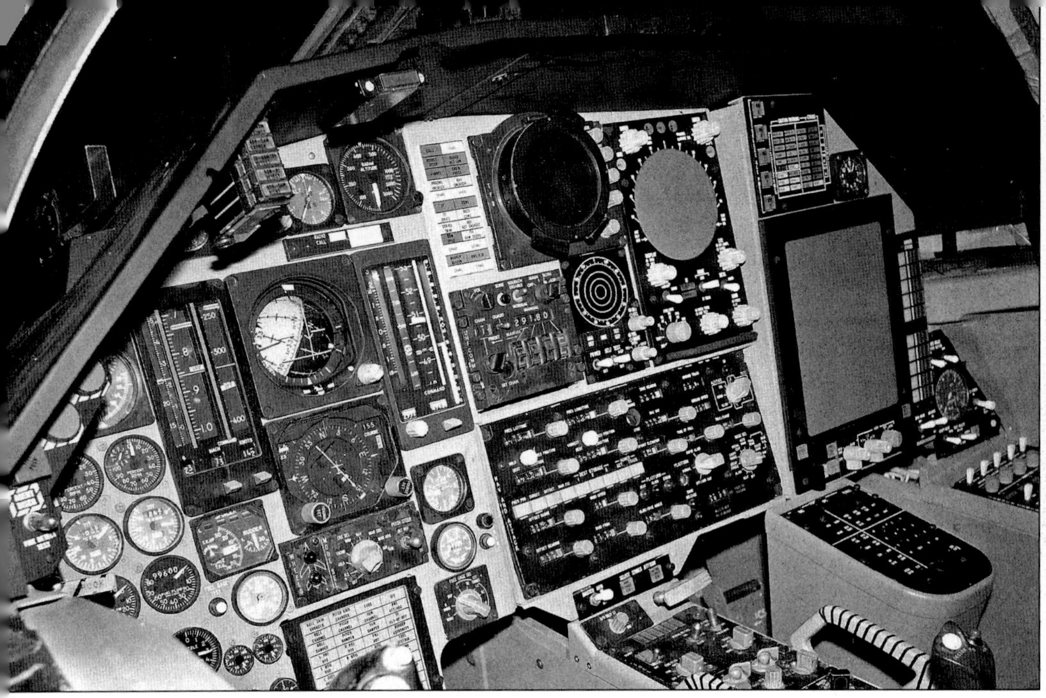

The EF-111A Raven cockpit contains a pilot station located on the left and the ECO station located on the right side. (Northrop Grumman History Center)

The pilot station has standard flight instruments located directly in front of the stick. ECM instruments are in the center instrument panel. (Northrop Grumman History Center)

The Raven's center panel, like the Aardvark, had fuel management and Terrain Following Radar (TFR) controls. It expanded to include ECM. (Northrop Grumman History Center).

The right side ECO console controls the jammer functions for the EF-111A. Control knobs adjust the signal frequency and strength. (Northrop Grumman History Center)

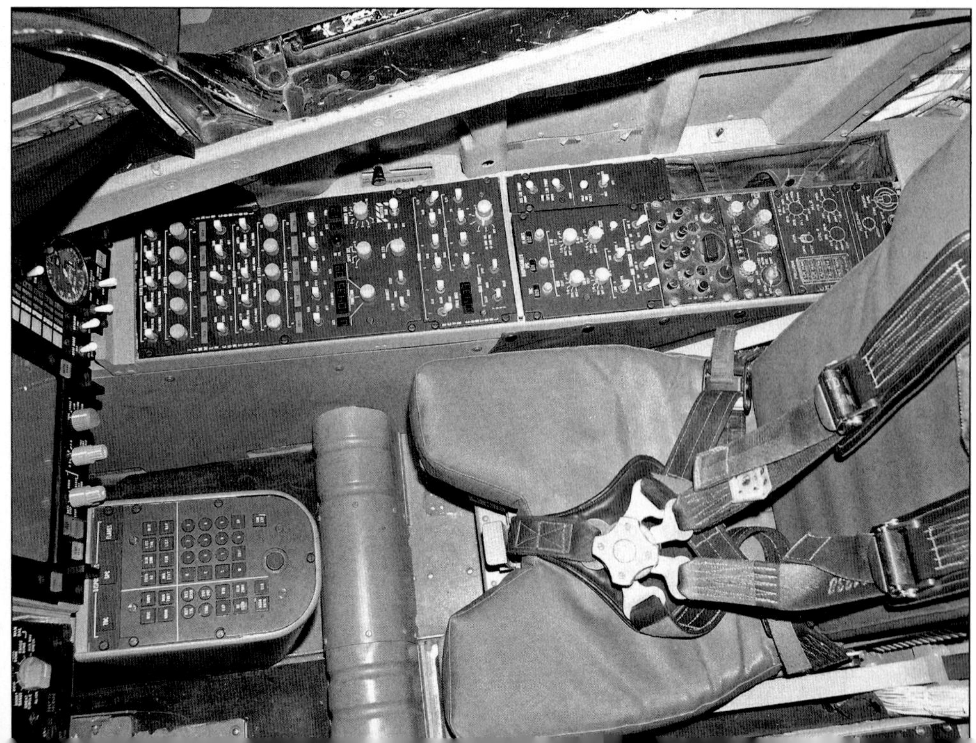

Grumman workers modify an existing F-111A Aardvark into a Raven ECM aircraft, EF-111A 66-049, at the plant in Calverton, New York, in June of 1981. (U.S. Air Force)

Once completed, EF-111A 66-040 rolls out of the hanger for a test flight. Grumman modified 42 of the F-111A aircraft into the EF-111A at the Calverton plant. (U.S. Air Force)

EF-111A 66-049 prepares for its first flight out of the Grumman facility in Calverton, New York in June of 1981. (Staff Sgt. Ernest H. Sealing / U.S. Air Force)

An EF-111A taxis between Korean air force A-37 Dragonfly and T-33 Shooting Star aircraft at Sachon Air Base in Korea in March of 1985. (Master Sgt. David Craft / USAF)

Unlike F-111 Aardvarks that wore Southeast Asia (SEA) camouflage, EF-111A Ravens were painted light gray and white. (Northrop Grumman History Center)

The left rotator glove assembly on the wing is tilted downward on an EF-111A aircraft to open the doors and allow full slat extension. (Northrop Grumman History Center)

The ALR-62 Track While Scan (TWS) antenna for the AN/ALQ-137 ECM system is located on the left side of an EF-111A Raven. (Northrop Grumman History Center)

A pneumatic ground cart duct is attached to an EF-111A. It blows compressed air to spin a turbine and engage a ring gear to start the engine. (Northrop Grumman History Center)

Prototype EF-111A 66-049 has red trim and a lightning motif while it undergoes tests at the Grumman Calverton facility in 1972. (Northrop Grumman History Center)

EF-111A 66-041, based in Upper Heyford, United Kingdom, lands during deployment to Aviano, Italy, in 1989. Raven paint schemes have light gray fins and tops. (USAF)

EF-111A 66-041 sports a light gray top paint with a white lower portion after conversion at Grumman's Calverton plant in New York. (Northrop Grumman History Center)

A gray and white EF-111A Raven (bottom) flies with an older model F-111 Aardvark (top) painted in the Southeast Asia (SEA) camouflage scheme that was used Vietnam. (USAF)

EF-111A 66-031 flies over mountains in the western United States during the 1970s. (Northrop Grumman History Center)

EF-111A 66-041 was the second Raven converted. It first flew as a test aircraft in March of 1977. (Northrop Grumman History Center)

EF-111A 66-051 blends into the clouds during an Air Force acceptance test flight over New York during the 1970s. (Northrop Grumman History Center)

EF-111A 66-051 flies a mission over the northern Bosnian Alps during Operation Deliberate Force in 1995. (Tana R. Hamilton / U.S. Air Force)

An EF-111A crew prepares for takeoff during Operation Desert Storm in January 2001. The radar station markings on the side of the fuselage indicate successful radar station jamming missions. In addition to the jamming missions, one unarmed EF-111A piloted by Capt. James Denton with ECO Capt. Brent "Brandini" Brandon was credited with an aircraft kill on 17 January 1991 when the crew outmaneuvered a pursuing Iraqi F-1 Mirage and forced it to crash into the ground. (Tech. Sgt. Fernando Serna / USAF)

A pair of EF-111A Ravens prepare for takeoff during Operation Desert Storm. Eighteen Raven ECM aircraft flew 900 sorties during the war with the loss of one aircraft. (USAF)

An EF-111A Raven from the 366th TFW (bottom) waits as an F-111F of the 48th TFW taxies to a runway during Operation Desert Shield. (Tech. Sgt. Rose Reynolds / USAF)

Ground crewmen work with EF-111A 66-6015 from the 429th Electronic Combat Squadron, 27th Fighter Wing, located at Cannon Air Force Base in New Mexico in February of 1997. The Raven was near the end of its EF-111A active service life as all Ravens were retired in 1998. The ventral canoe radome is mounted on the lower fuselage between the rear of the nose landing gear and the front of the engine inlet. (Tech. Sgt. John McDowell / U.S. Air Force)

An EF-111A flies during Theater Force Employment Exercise IV. Ravens have an ECM bay with a radome mounted on the lower front of the fuselage. (Sullivan / U.S. Air Force)

The various transmitters for Bands 4, 7, 8, and 9 are embedded in the lower fuselage of an EF-111A with a canoe radome on the bottom. (Northrop Grumman History Center)

An EF-111A of the 42nd ECS fires afterburners during a takeoff from a United Kingdom airfield in late 1991. It and three others deployed to Turkey. (Cynthia Alderson / USAF)

An EF-111A Raven from the 42nd ECS turns onto a taxiway at a United Kingdom airfield during exercise Display Determination '89. (Staff Sgt. David S. Nolan / U.S. Air Force)

Before an airstrike in Bosnia, Senior Airman Odam logs chaff flares for an EF-111A at Royal Air Force (RAF) Lakenheath in 1995. (Tech. Sgt. Bruce Sherwood / USAF)

A ground crew performs pre-flight checks before an ECM mission over Bosnia on an EF-111A from the 429th ECS in Italy in 1995. (Tech. Sgt. Bruce Sherwood / USAF)

A KC-135 Stratotanker refuels an EF-111A Raven of the 429th ECS deployed in Italy and assigned to enforce the no-fly zone over Bosnia in 1995. (Tana R. Hamilton / USAF)

Most EF-111A ended service at Davis-Monthan AFB in the 309th Aerospace Maintenance and Regeneration Group (AMARG) facility, often called "The Boneyard." (Del Loughery)

EA-18G Growler

An EA-18G Growler approaches a runway with a legacy AN/ALQ-99 low-band jammer pod located underneath the fuselage centerline in New York on 17 November 2011. The current AN/ALQ-99 jammer pods are slated for a redesign that is designated as the Next Generation Jammer (NGJ). AN/ALQ-218 pods are mounted on each wingtip for the jammer-receiver system.

The EA-18G Growler replaced the U.S. Navy's EA-6B Prowler. For the foreseeable future, the Navy is the only U.S. military service with a dedicated ECM aircraft for the jammer mission. The EA-18G Growler is the heir apparent for the Navy's continued involvement in ECM aircraft. The EA-18G is manufactured by Boeing in St. Louis, Missouri, and the design is based on Boeing's F/A-18 Super Hornet airframe. Currently the EA-18G carries the Northrop Grumman-manufactured AN/ALQ-99 legacy jammer pods. These were previously carried by the EA-6B aircraft, which will be retired by 2019. (Ken Neubeck)

A Navy two-seat F/A-18F Super Hornet prepares to touch down in New York. It is the base airframe for the EA-18G. (Ken Neubeck)

An EA-18G Growler approaches to land in New York. The EA-18G Growler has wingtip pods and jammer pods mounted on the fuselage. (Ken Neubeck)

An F/A-18F is chocked on a hardstand. F/A-18F basic airframe used for the EA-18G was first introduced into Navy service in 1999. (Ken Neubeck)

The EA-18G is parked on a hardstand in New York. The EA-18G was first introduced into service in 2007. (Ken Neubeck)

The front EA-18G Growler cockpit position is for the pilot, who performs no ECM tasks. The control panel features standard flight control and gages, along with multi-function glass displays similar to the F/A-18 Super Hornet. The ejection seat is the modern SJU-17 (Mk. 14) Navy Aircrew Ejection Seat (NACES) made by Martin Baker. (Jamie Hunter)

The rear station of the EA-18G Growler is for the ECM operator and includes four multi-function displays used to track and jam signals. The ECM operator's station has no flight controls. This separation ensures a complete autonomy of duties between the pilot and ECM operator. The ECM operator also has the same model ejection seat. (Jamie Hunter)

Lt. Ryan T. Mattson, VAQ-129 instructor pilot, (left) and his father, Capt. Stephen D. Mattson (right), prepare to fly an EA-18G on 20 April 2012. (Nardal Gervacio / U.S. Navy)

The EA-18G pilot opens a door panel below the wing to remove personal gear. A smaller stowage panel door is ajar toward the front of the aircraft in New York. (Ken Neubeck)

Vice Adm. Allen G. Myers (left) sits in the ECM operator seat while Lt. Cmdr. Kevin Inabit of VAQ-129 (right) pilots during a flight in Washington in 2011. (Nardal Gervacio / Navy)

Capt. Chris Rodeman, NAF Misawa commanding officer, (left) gets advice from Lt. j.g. Scott Chuda (right) before a Growler flight in Japan in 2013. (Kenneth G. Takada / Navy)

Formation and navigation lights are illuminated on the fuselage and tail of an EA-18G from VAQ-129 on the USS *Carl Vinson* on 26 February 2013. (Timothy A. Hazel / Navy)

With its holdback bar attached to the shuttle, an EA-18G from VAQ-141 prepares for a catapult-assisted launch from the USS *George H.W. Bush*. (Nicholas Hall / U.S. Navy)

Navy safety observer Riley Hart (left) watches carefully as an EA-18G Growler from VAQ-129 crouches for a catapult-assisted launch from the USS *John Stennis.* The catapult control officer prepares the catapult system from the Integrated Catapult Control Station (ICCS or "The Bubble") below the deck on 30 March 2011. (Kevin Murphy / U.S. Navy)

Smoke billows from the exhaust of an EA-18G aircraft as the pilot starts its engines to leave Republic Airport in New York after an educational visit. (Ken Neubeck)

An EA-18G Growler aircraft from the VAQ-132 "Scorpions" taxis to the main runway at Republic Airport in New York on 18 November 2011. (Ken Neubeck)

An EA-18G from the VAQ-141 "Shadowhawks" lands on the USS *George Washington* during operations between the United States and Australia. (Mathew Riggs / U.S. Navy)

An EA-18G Grower from the VAQ-141 "Shadowhawks" fires its afterburners during a launch from the aircraft carrier USS *George H.W. Bush*. (Kevin J. Steinberg/U.S. Navy)

Five EA-18G Growlers from VAQ-132 return to NAS Whidbey Island in Washington after an eight-month deployment in Libya and Iraq on 9 July 2011. (Nardel Gervacio / Navy)

Growler EA-18G 166938 from VAQ-138 taxis down the runway at Nellis AFB in Nevada during exercise Red Flag 11-3 on 8 March 2011. (Staff Sgt. William P. Coleman / USAF)

EA-18G 166894 from the VAQ-132 “Scorpions” prepares for a mission. The color paint on the tail indicates that its the wing commander’s aircraft. (Nardel Gervacio / U.S. Navy)

A pair of EA-18G from the VAQ-132 “Scorpions” fly over New York after a deployment in Libya as part of Operation Odyssey Dawn in October of 2011. (Ken Neubeck)

An EA-18G gets fuel at Republic Airport in New York on 10 June 2010. It is equipped with the full complement of three jammer pods. It carries one high-band jammer pod under each wing, and a low-band pod under the centerline. (Ken Neubeck)

The centerline low-band jammer pod is elongated and heavier than the high-band pods because has a broader signal pattern and more internal components. (Ken Neubeck)

An EA-18G is parked at Republic Airport in New York on 17 November 2011. It carries a low-band pod under the fuselage centerline. It has no high-band pods. (Ken Neubeck)

A high-band pod is mounted on the outside pylon of an EA-18G in New York in 2010. Smaller high-band pods are more directional than the low-band pod and are mounted on the outside wing pylon for better antenna pattern for high frequencies. (Ken Neubeck)

A high-band pod is mounted on a Growler. These pods are more directional and coverage is impacted on the inboard side when external fuel cells are attached. (Ken Neubeck)

In the nose of each pod has a ram-air turbine system used to generate power for the pod's electronics. Electronics are cooled by air inducted through the hole. (Ken Neubeck)

Antennas are located on the outside panel in each pod. Warning notes are provided on the pods to ground crew during maintenance. (Ken Neubeck)

Warning notes, along with notes to indicate antenna access panels are provided for ground workers along the entire length of the centerline pod. (Ken Neubeck)

An EA-18G equipped with a low-band pod has its taxi lights illuminated on the two tail fins at Nellis Air Force Base in Nevada on 7 June 2011. (Jamie Hunter)

An EA-18G from the VAQ-140 "Patriots" launches from the USS *Harry S. Truman* during a night exercise in the Atlantic Ocean in September 2015. (J.R. Pacheco / U.S. Navy)

Maj. Benedict G. Buerke trains on the AGM-88 HARM mounted under an F/A-18 Hornet's wing on 31 March 2011. EA-18G also use the missile. (Cpl. Ben R. Reynolds / USMC)

An EA-18G aircraft from the VAQ-141 "Shadowhawks" moves into position for a catapult-assisted launch from the aircraft carrier USS *George H.W. Bush*. (Billy Ho / U.S. Navy)

An EA-18G Growler from the VAQ-129 "Vikings" squadron takes off at NAF El Centro in California during a training exercise on 29 October 2008. (Rialyn Rodrigo / U.S. Navy)

The tailhook of an EA-18G from the VAQ-141 "Shadowhawks" grabs the arresting cable on the deck of the carrier USS *George Washington*. (Paul Kelly / U.S. Navy)

An EA-18G Growler from the VAQ-141 "Shadowhawks" launches from the carrier deck of the USS *George W. Bush* on 23 October 2010. (Betsy Lynn Knapper / U.S. Navy)

An EA-18G Growler from the VAQ-141 "Shadowhawks" touches down on the deck of the carrier USS *George Washington* in the Philippine Sea in 2012. (William Pittman / Navy)

An EA-18G test aircraft releases decoy flares from the AN/ALE-47 countermeasures dispenser over the Navy test range at China Lake in California on 24 October 2013. New production EA-18G Growlers are being deployed into field service and is the heir apparent for the electronic jammer aircraft role in the early 21st Century as the Navy gradually retires EA-6B Prowlers. While the Air Force has integrated some ECM and stealth technology into their aircraft, the Air Force will not pursue an ECM-only aircraft. The EA-18G, however, will support the Navy's ECM mission for many years to come. (Jamie Hunter)